Marxist Political Economy as a Science by Hajime Kawakami
2011 Prism Key Press / www.prismkeypress.com

Marxist Political Economy as a Science
Hajime Kawakami

Contents

The highest task of humanity is to comprehend this objective logic of economic evolution (the evolution of social life) in its general and fundamental features, so that it may be possible to adapt to it one's social consciousness and the consciousness of the advanced classes of all capitalist countries in as definite, clear and critical a fashion as possible.

Lenin: *Materialism and Empiro-Criticism*

1. The Mission of Science

–The Essence and Phenomenal Forms of Things–

A general election was recently held in Japan. According to a survey by the Ministry of the Police, of the total 9.58 million valid votes cast, proletarian political groups received 470,000 votes. This means that ninety-five percent of the citizens lent their support to bourgeois parties, with only five percent supporting proletarian ones

Here we see a completely reverse reflection of the actual state of society. Political parties representing the wealthy and the large landowners, who are a small minority – roughly five percent of the population – were able to obtain an overwhelming ninety-five percent of the votes. Meanwhile, political parties representing those who are not wealthy, who make up ninety-five percent of the population, only managed to obtain five percent of the total votes. Such a result arises from the masses of people having an upside-down awareness of the essence of reality.

The destiny of any science is to run counter to the common sense that prevails. There would be little need for science if the layman were able to understand an object through simple thought that does not rely upon scientific examination. But what the layman thinks is often the reverse of the essence of reality, making it necessary to overturn such thought through the scientific consideration of a given matter. This means that science is destined to struggle against common sense, so that only a courageous person is capable of becoming a true scientist.

Let me offer one, easy-to-understand example. The earth orbits the sun while rotating on its own axis-not the reverse-but to our eyes it seems that the sun is instead orbiting the earth. What is reflected in our field of vision is the sun rising every morning in the east, then setting in the west in the evening. Given this appearance, no one had thought that the planet we inhabit is itself in motion. The consciousness of those who grasped this phenomenal form, which is the form that appears to the naked eye, was in fact the reverse of the essence (or truth) of the matter. This was the source of their error, which acted as an enormous barrier to cultural development

Here is another example. A man who consumes a few alcoholic drinks may turn red in the face, and even in winter feel warm enough to take off an article of clothing or open the window. Another person, seeing this, would think that the man's body is warm. In fact, however, a person's body temperature will decrease after consuming alcohol, as verified by tests carried out by specialists, and which we can also simply demonstrate using a thermometer. The drinker feels warm because the alcohol dulls the nerves that perceive cold. There are nerves in our skin that perceive cold, and these nerves play the role of sentries that communicate to the brain the fact that cold air has penetrated. But when a person has a few drinks, these sentries gradually start to doze off, so that when the body's temperature gradually drops, and the order to do something about this does not reach the hub of the brain, the drinker gradually feels warmer even though the body's temperature is dropping, due to a weakened perception of cold.

Science not only teaches us that the essence of a matter can be the reverse of its phenomenal form, but also explains why the essence comes to have a phenomenal form that is its opposite. In addition to teaching us that the earth, not the sun, is in motion, it also explains why our eyes perceive that the earth is in motion rather than the sun. (This is the dialectical unity between essence and phenomenal form.)

The people who are leading establishment political parties may think that they are striving for the interests of the country and the welfare of the people, but we cannot judge such people according to what they think of themselves. A person who thinks himself good is not always in fact so. And even a drinker who thinks himself warm will in fact have a falling body temperature. There may also be many people who see establishment parties as promoting their own interests. But this is no different from a drinker appearing to be warm or the sun appearing to rise in the east and set in the west.

If someone says that the sun is in motion, it certainly does appear to be the case in one's eyes, and this is an explanation that people found easy to accept, just like the idea that drinking warms a person up.

This makes it easy to deceive people. It is easy to lie, whereas a true explanation is complicated. And when those with the power of the state and money on their side are the ones freely spreading lies, it is certainly no easy task for a person lacking this power to reveal the truth to others under conditions where extreme limitations are placed on personal speech.

These are the reasons that give rise to a situation where establishment parties, despite representing the wealthy and the large landowners – who make up around five percent of the population – were able to gain an overwhelming ninety-five percent of the votes in the recent general election, whereas the parties of those without property – about ninety-five percent of the population – only managed to win five percent. The establishment parties base their constituency on the dormant masses of people, who are prey to an illusion wherein reality is perceived in a form that is the opposite of the truth.

Whether it be a natural or social phenomenon, there are often illusions that are the reverse of the essence of a matter. In terms of overturning such illusions alone, science is revolutionary. It was a revolutionary task to overturn the idea that sun orbits the earth in order to recognize that in fact it is the earth that is in motion. And what Marx achieved in the realm of political economy was also a revolutionary task.

Established political power, of course, always has an interest in taking advantage of people's illusions-making use of the illusions of those who think in terms that are the opposite of the truth of a matter-and this power has a keen interest in maintaining and promoting this state of affairs. Every instrument of propaganda available is mobilized to this end, including schools, youth groups, religious organizations, newspapers, magazines, books, and so on. In the recent general election as well, numerous members of establishment parties delivered one speech after speech, from an infinite number of rostrums, to achieve this purpose. As pointed out in book one of *Musansha seiji kyōtei* [A Course on Proletarian Politics], "when the causes of troubles are concealed, and when people are unable to combat these causes, there is no way for those suffering from these troubles to use their own power to change the situation." And this suits the ruling class just fine. Established power, therefore, is always hostile to new science. In the recent election, as the reader knows from newspaper articles, extreme limits were placed on our speech, which I happened to experience first-hand on a number of occasions.

New science can only develop itself by engaging in a struggle against powerful established forces. Scholars who in the past declared that the earth, not the sun, is in motion, were thrown into prison. There were scholars who called for the abandonment of the idea that the planet we are standing upon does not move, and they explained that the earth is capable of movement and is in fact in motion, which brought fear to the ruling class at the time. So those who today seek to do something similar regarding social phenomena – which is a domain where a direct negotiation takes place between people's life interests – naturally generate a barrage of criticism, slander, misunderstanding, pressure and the like, and one needs to be prepared for this. As Marx writes:

Free scientific inquiry [in the domain of political economy] does not merely meet the same enemies as in all other domains. The peculiar nature of the material it deals with summons into the fray on the opposing side the most violent,

11

sordid and malignant of the passions of the human breast, the Furiers of private interest.[1]

The "ultimate aim" of *Capital*, as Marx notes in this same preface to the first German edition, is to "lay bare the economic law of motion of modern society." The economic law of motion referred to is similar to the physical law that governs the earth's motion. Just as people once believed that the planet they inhabit does not move at all, so people today are resigned to the idea that the modern social organization they live within is eternal and unchangeable. But *Capital* teaches us that what appears to be unchanging is capable of movement and is in fact in motion. This reveals the "special laws that regulate the origin, existence, development, death of a given social organism and its replacement by another and higher one." *Capital* is a book that awakens the oppressed from their resignation based upon illusions, reveals to them the causes of their troubles, while also indicating the possibility and the path to sweep these troubles aside, thereby instilling in them hope and a direction for their own self-liberation. In a word, *Capital* is a crucially important textbook for revolution[2].

There is no need to speak further here about the outcome of human action led by scientific knowledge. Our predecessors though that *thunder* was the fearsome anger of the gods. But today, as the outcome of scientific research on electricity, not only do we know that this is nothing more than the sound of electricity being discharged in the sky, but this once fearsome specter has also been transformed into something docile and obedient that is utilized for telegrams, telephones, lights, trains, etc, and for the powering of fans in the summer to generate wind or electric stoves to heat rooms in winter. In short, the progress of science has turned what had been a most fearsome thing into its opposite: something that is highly useful. Thus, for political economy to merit the name "science," it must be able to furnish guiding principles to transform the powerless masses of people, who are submerged in their troubles, into the opposite-strong people who enjoy a happy life. And Marx's book does indeed have this value.

Notes

1. Karl Marx, *Capital*, vol. 1, trans. Ben Fowkes (London: Penguin Books, 1976) 92.

2. The word "revolution" (*kakumei*) was removed by censors and replaced with "xx."

2. Revealing the Particular Laws of Motion within Capitalist Society

"Vulgar economists" are those economists who do not deserve to be considered scholars. They are so named because they are on the same level as a vulgar person. Completely attached to phenomenal forms, they do not seek to grasp the essence of a matter. For example, when it appears that the sun is moving from east to west, this is the superficial, phenomenal form reflected in our eyes, but a vulgar scholar does no more than describe this phenomenal form – making it sound plausible through the use of scholarly terminology – taking not a single step beyond this level. Their "scholarship" amounts to nothing more than lining up items of common sense. Given this, even if their ideas are easy to understand, they are without usefulness. Such scholars, needless to say, exist in droves in various countries, such as the historical school of political economy in Germany or the psychological school in Austria. These two schools are essentially identical and yet superficially opposed to each other, just like the Seiyukai and Minsei parties in Japan compete with each other despite both being bourgeois political parties. The methods of the two schools are different, with the former listing up facts instead of elucidating laws (thereby missing the forest for the trees), while the latter offers up abstract laws in place of the particular laws of capitalist society. Still, both schools are identical in terms of rejecting the special laws of capitalist society. Thus, just as the founder of the historical school, Wilhelm Roscher, discovers the primitive accumulation of capital among primitive peoples catching fish using their bare hands, Bohm-Bawerk, the head of the psychological school, sees the characteristics of capitalist production in the example of water being obtained through a *bamboo pipe*. Both schools are identical in terms of turning the concept of capital into something general, abstract and external.

Marx, contrary to this, seeks in *Capital* to reveal the particular historical laws of modern society (capitalist society). Regarding this, we can find the following passage in the post face to the second edition of that work:

It will be said...that the general laws of economic life are one and the same, no matter whether they are applied to the present or the past. But this is exactly what Marx denies. According to him, such abstract laws do not exist...On the contrary, in his opinion, every historical period possesses its own laws...As soon as life has passed through a given period of development, and is passing over from one given stage to another, it begins to be subject also to other laws...The old economists misunderstood the nature of economic laws when the likened them to the laws of physics and chemistry. A more thorough analysis of the phenomena shows that social organism differ among themselves as fundamentally as plants or animals. [3]

For example, the various social organisms of slave-based society, feudal society and capitalist society are not equivalent to the childhood, adolescence and adulthood of the *same* living organism. Rather, they are like three completely different living organisms, such as the differences between a snake, a dog and a human being. A snake does not develop into a dog, just as a dog does not develop into a person. Each is rather a *fundamentally distinct* living organism, and therefore "one and the same phenomenon" will even "fall under quite different laws in consequence of the different general structure of these organisms, the variations of their individual organs, and the different conditions in which those organs function."[4]

In his introduction to *Foundations of the Critique of Political Economy* (*Grundrisse*), Marx notes:

Although it is true, therefore, that the categories of bourgeois economics possess a truth for all other forms of society, this is to be taken only with a grain of salt. They can contain them in a developed, or stunted, or caricatured form etc., but always with an essential difference. The so-called historical presentation of development is founded, as a rule, on the fact that the latest form regards the previous ones as steps leading up to itself, and, since it is only rarely and only under quite specific conditions able to criticize itself-leaving aside, of course, the historical periods which appear to themselves as times of decadence-it always conceives them one-sidedly.[5]

As long as a person believes that the society he is living in is the ultimate form of society, the social forms of the past will appear to be nothing more than a means of reaching this goal, thus losing their own independent existence. The past forms do not seem to be in opposition to this final form but rather subordinate to it. This accordingly rules out a comprehensive understanding of the past forms. This is the reason why "the latest form regards the previous ones as steps leading up to itself" and "always conceives them one-sidedly."

Thus, as Marx explains in *Capital*, we cannot overlook the specific laws of capitalist society.

Marx thus says: "What I have to examine in this work is the capitalist mode of production, and the relations of production and forms of intercourse [*Verkehrsverhältnisse*] that correspond to it."[6] The mechanism of capitalist production, as I shall explain in a moment, is the specific mechanism of production wherein capitalists employing numerous wage workers, themselves take possession of all of the products produced by the labor of workers in return for paying them a wage. A society where such a mechanism has come to be the dominant power is a society where "the capitalist mode of production prevails," or simply put: a capitalist society.

Capital elucidates the specific laws of such a capitalist society. We must note, however, that a pure capitalist society has yet to exist, and could not exist. Within actual society, therefore –

16

regardless of the degree to which capitalist development has been achieved – there will always be some part of it that is made up of past social forms, "maintained as a remnant that was not overcome." And, needless to say, the theory in Marx's *Capital* cannot be directly applied to these remnants of the past.

The population of rural agricultural villages in Japan, according to a 1921 survey, accounts for 48.2 percent of the entire population, but most of the agriculture in these villages is still not carried out under capitalistic production methods, and is therefore not capitalist agriculture. This means that the theory of capitalist ground rent explained by Marx in *Capital* is not applicable to the farm rent (kosakuryo) paid by tenant farmers in Japan. This was already effectively indicated by Murayama Toshirō[7] in *Marxism and Problems of Agriculture*, where he writes:

In Japan, tenant rent paid in kind is the dominant form of ground rent. There has arisen the peculiar phenomenon of mistaken arguments being raised by those who have not seriously considered this fact and instead attempt to apply the schema of Marx's theory of ground rent to tenant rent in Japan. This can be seen, for example, in the work of Motoyuki Takabatake, Sentarō Kitaura and others.

Notes

3. *Capital* vol. 1, 101.

4. Ibid.

5. Karl Marx, *Grundrisse*, trans. Ben Fowkes (London: Penguin Books, 1973), 106.

6. *Capital* vol. 1, 90.

7. Names are listed with the family name second, which is the

opposite of the custom in Japan.

3. Materialist Starting Point (External Phenomena as the Starting Point of Investigation)

–The Commodity as the "Cell" of Capitalist Society–

Every effort is being made, from every direction, to prevent those submerged in troubles from locating their source. If we resist such confusion and identify the truth, we will discover that the source of problems is not that the area of land is too small, the population too large, foreigners too domineering, nor that we are too ignorant. This source can rather be traced to the organization of society. And we can also discover that if we want to eliminate these problems, we have no choice but to fundamentally revolutionize[8] this organization.

First and foremost, we need to identify as accurately as possible what exactly is the structure of the world we inhabit and the direction it is headed.

If we are to accurately know this, our method of cognition must be correct. And this correct method is the materialist dialectic. Here I would like to first say a word about the standpoint of materialism, and then deal briefly with the nature of a dialectical understanding.

What is the economic mechanism of the world we are living in? If we were idealists, when examining this problem we would likely close our eyes and cross our arms. But we are materialists and believe that only a materialist method of cognition will provide us with a correct understanding – an objective, scientific understanding – so we begin by opening our eyes and looking around at the phenomena that surround us. There is no way to avoid being taken in by the phenomenal forms that appear in our eyes, but our investigation must begin with

19

these external phenomena as they appear to our eyes. Our starting point must be what is reflected in everyone's eyes, what everyone would agree that they also see; the clear and indisputable facts that would satisfy anyone. Because this is our point of departure, our research is able to be reliable. If, instead, we were to start with what we are thinking inside our own hearts, even if this is thought to be correct, it would be doubtful whether this is indeed the case. And if one sought to convince others, some further proof would be necessary, meaning that the argument would wander hither and thither, with the entire investigation ending in error.

So we start from the clear facts that can satisfy everyone. In the post face to the second German edition of *Capital*, there is the following passage regarding this that Marx cited from a review of his book:

Marx treats the social movement as a process of natural history, governed by laws not only independent of human will, consciousness and intelligence, but rather, on the contrary, determining that will, consciousness and intelligence.... If in the history of civilization the conscious element plays a part so subordinate, then it is self-evident that a critical inquiry whose subject-matter is civilization, can, less than anything else, have for its basis any form of, or any result of, consciousness. That is to say, that not the idea, but the material phenomenon alone can serve as its starting-point.[9]

Our economic lives appear in people's eyes as a clear fact. – In order to live, we know that labor must be directed towards the external, natural world in order to create food, clothing and shelter, along with other things. In a letter to Ludwig Kugelmann, written on July 11, 1868, Marx writes:

Every child knows that any nation that stopped working, not for a year, but let us say, just for a few weeks, would perish. And every child knows, too, that the amounts of products corresponding to the differing amounts of needs demand differing and quantitatively determined amounts of society's aggregate labor. It is **self-evident** that this *necessity* of the *distribution* of social labor in specific proportions is certainly not abolished by the *specific form* of social production; it can only change *its form of manifestation.* Natural laws cannot be abolished at all. The only thing that can change, under historically differing conditions, is the *form* in which those laws assert themselves.[10]

This means that human beings, in order to live, have no choice but to create things through labor. So if such labor were to cease, even less than a year, every human being would also cease to exist. Also, because human beings have many needs, the labor of society as a whole must be distributed to each of the production sectors at a given proportion in order to produce the various products that correspond to these needs. This is a natural law that applies to any organization of society as long as there are people living upon the planet. The change that occurs between different social organizations involves the type of phenomenal form that this natural law is manifested within. All of this, in the words of Marx, is "self-evident" even to a child. Likewise in today's society, it is evident that labor must be continually expended to produce the things needed to maintain our lives. And this aggregate labor of society must be distributed to the production sectors engaged in iron-making, shipbuilding, spinning, textile production, agriculture, mining, etc. However, all of the products produced are produced as commodities, so that social production is carried out through the method of commodity production. This means that in society today the natural law mentioned above is manifested specifically as commodity production.

This is why Marx, at the very beginning of *Capital*, writes: "The wealth of societies in which the capitalist mode of production prevails appears as an 'immense collection of commodities'; the individual commodity appears as its elementary form."

This is also an external phenomenon that appears in the

21

eyes of anyone living today. Even a child would be aware of this. Instead of producing what is needed for one's own family, a person is able to obtain wealth, via payment, that is produced by other people (society) and comes in diverse varieties and infinite quantities. But everything is in the commodity form. This is the reason why "the wealth of societies in which the capitalist mode of production prevails appears as an immense collection of commodities." It is called an "immense collection of commodities" because any and every thing becomes a commodity, so that anything can be obtained if money is paid and every product without exception is available as a commodity. But this has not always been the case. When commodity production had yet to develop sufficiently, there was no one to buy from even when a person wished to buy something, leaving that person no choice but to produce for their own consumption. But as the capitalist method of production gradually took hold, large numbers of products came to assume the commodity form. This can be understood if we journey from the countryside of Japan to the city. In rural areas, where capitalist production has yet to fully penetrate, one can get along to a greater extent without money. But in the city, where capitalist production has established itself as the prevailing force, one has to pay money for nearly everything. And anything can be bought, provided that money is paid. If we stroll over to the nearest department store, we will be afforded a first-hand view of a "collection of commodities."

In this manner, the wealth of capitalist society appears as an "immense collection of commodities" and the individual commodity is the cell of this social wealth. Just as the human body is composed of an infinite number of individual cells, the wealth of capitalist society is composed of an infinite number of individual commodities. This is also something that can be clearly seen by anyone. If a person heads to the city to obtain something he wants, the item will have a price tag attached to it, or if not a sales clerk will tell the person how much it costs. At any rate, it is immediately clear to anyone that every item on the store's shelves is a commodity. This is why Marx says that the individual commodity appears as the "elementary form" of the wealth of capitalist society. Here "appears" means that it is reflected in our eyes as an external phenomenon or phenomenal form. This phenomenal form is the starting point of *Capital*.

The points above can also be explained in the following manner.

The problem we are dealing with is the nature of the economic mechanism of the world we live in. If we take a glance at this world, we can see the economic relations between people in present-day society and discover the exchange relation between one commodity and another. As long as we are unable to survive without acquiring what other people produce, everything that people (particularly those living in cities) require – which includes their rice, miso, soy sauce, fish, meat, milk, vegetables, and other food, as well as their clothes, utensils, newspapers, pens, ink, and so on – is produced by other people. Things that are either stolen or received for free constitute a rare exception to this. As a rule, things must be purchased for a given amount of money. Because of this there are infinite exchange relations established between people, and the total of these exchange relations constitutes the economic structure of society today.

Seen from this standpoint, therefore, the economic structure of capitalist society appears as an "immense accumulation" of exchange relations, and it could be said that the individual exchange relation is the "elementary form" of this. Therefore, according to Lenin, commodity exchange is the "simplest, most ordinary and fundamental, most common and everyday *relation* of bourgeois (commodity) society, a relation encountered billions of times"; in a word: the "cell" of capitalist society.

In short, our investigation should not begin right away with complex things. Rather, we must move progressively from the simplest things to ascend towards the more complex. This is why Marx says: "Our investigation therefore begins with the analysis of the commodity."

Notes

8. The words "fundamentally" (*konponteki*) and "revolutionize" (*henkaku*) were removed by censors and each replaced with "xx."

9. *Capital* vol. 1, 101.

10. Karl Marx, *MECW* vol. 43 (New York: International Publishers, 1988), 68.

4. A Dialectical Grasp

–The Locus of Every Contradiction of Capitalist Society is Uncovered through the Analysis of the Commodity–

Next, I would like to say a word about the manner of a dialectical grasp.

In modern society, needless to say, a variety of contradictions appear from various directions. As one example I can point to the following newspaper article I happened to come across in today's *Osaka Mainichi* (March 2, 1928):

Electrical companies are apparently suffering from an excess of electricity, claiming that they may be unable to continue production unless a limitation is placed on production. But this is outrageous! Where exactly is this supposed excess electricity? The electricity, said to be in excess, is apparently being conveyed through wires, but it fails to reach the most basic users. Of all the types of fuel, electric heating is the most convenient, but it is the most expensive and thus not commonly used. We have to make do with very dim electric lighting. In mountainous parts of the country, which are full of tunnels, the trains still burn coal that generates a great amount of soot. Even though railways are built for political reasons where there are neither passengers nor belongings to transport, we have yet to hear someone talk about using the excess electricity to electrify the railways. We often hear commercials about the electrification of households and a modern lifestyle, which everyone knows is convenient, but this is not affordable. Yet, what could make less sense than the idea that there is a problem of excess electricity or the call to reduce production? This would mean that the machines built will not be used, with water falling from waterfalls while these machines merely watch it flow out to the ocean. What a waste! And the electrical companies, despite this waste, still sell electricity at a high price, thus impeding the realization of a modern lifestyle, all the while talking about excess electricity.

Here we have an example of why many people fall into poverty in the midst of wealth. Electricity is not reaching the general consumer. And yet the electrical companies are struggling because of an excess of electricity. If this is not a contradiction what should we call it? Moreover, to resolve this contradiction, electrical companies are seeking to limit production. This means that the facility that we went to the trouble of building will not be used. And without this curtailment it is said that the companies may be unable to continue production. Under the system of present-day society, this contradiction can only be solved by limiting production-i.e. by curtailing productive power and deliberately preventing an increase in the production of wealth. This clearly signifies a clash between productive power and the relations of production, indicating that the current relations of production (the social relations or mechanism of society) have already become a barrier to the further development of productive power.

How did such a contradiction arise? Lenin writes:

The splitting of a single whole and the cognition of its contradictory parts is the *essence* of dialectics...The identity of opposites (it would be more correct, perhaps, to say their "unity"...) is the recognition (discovery) of the contradictory, *mutually exclusive*, opposite tendencies of *all* phenomena and processes of nature (*including* mind and society). The condition for the knowledge of all processes of the world in their "*self-movement*," in their spontaneous development, in their real life, is the knowledge of them as a unity of opposites.[11]

And he adds:

In his *Capital*, Marx first analyzes the simplest, most ordinary and fundamental, most common and everyday *relation* of bourgeois (commodity) society, a relation encountered billions of times, viz. the exchange of commodities. In this very simple phenomenon (in this "cell" of bourgeois society) analysis reveals *all* the contradictions (or the germs of *all* the contradictions) of modern society. The subsequent exposition shows us the development (*both* growth *and* movement) of these contradictions and of this society in the ? [summation] of its individual parts, from its beginning to its end.[12]

This simple presentation by Lenin adequately expresses the nature of a dialectical grasp within political economy.

The commodity as the "elementary form" of wealth in capitalist society – and the commodity as the "cell" that is the fundamental constituent part of capitalist society – must be analyzed first, as noted earlier, but this analysis seeks to split the commodity qua unified thing into its contradictory parts.

Once it is discovered that the commodity contains a contradiction, the movement of the commodity can be grasped as *self-movement*, because a contradiction is the origin of movement. Here self-movement means that this is not movement dependent upon something else, but rather movement that possesses a motive force of its own. It is only by grasping the movement of a thing in this manner that we can first come to a fundamental understanding. This is because if the motive force of movement does not exist within the thing itself, but rather within something else, our study would have to further trace this other thing, and thus would to attain its ultimate aim.

The fundamental opposites within the commodity are use-value and value. A commodity must be a useful thing for people other than its owner, which is to say that it must be a use-value-or material wealth. At the same time, every commodity has a certain price, in terms of being such-and-such Japanese *sen* worth of gold. The value encompassed in the commodity is expressed in money. So the commodity is dualistic – a use-value on the one hand and a value on the other.

A use-value is the outcome of productive power but value

is the expression of the relations of production. A commodity is the unity of opposites (use-value and value), and the commodity encompasses this fundamental contradiction. Encompassed within the commodity is every contradiction (and the germ of every contradiction) of commodity-production society – and of capitalist society as the highest development of commodity production.

The fact that electrical companies must limit production, even though electricity is being delivered to those who require it, is the result of electricity being produced as a commodity. If this is provided above a certain point, the unity of use-value and value is destroyed. Underlying this is a fundamental cause, which is the collision between the productive power of capitalist society and the relations of production within it.

In his preface to *A Contribution to the Critique of Political Economy*, Marx offers the general conclusion of his research:

In the social production of their existence, men inevitably enter into definite relations, which are independent of their will, namely relations of production appropriate to a given stage in the development of their material forces of production. The totality of these relations of production constitutes the economic structure of society, the real foundation, on which arises a legal and political superstructure and to which correspond definite forms of social consciousness. The mode of production of material life conditions the general process of social, political and intellectual life. It is not the consciousness of men that determines their existence, but their social existence that determines their consciousness. At a certain stage of development, the material productive forces of society come into conflict with the existing relations of production or – this merely expresses the same thing in legal terms – with the property relations within the framework of which they have operated hitherto. From forms of development of the productive forces these relations turn into their fetters. Then begins an era of social revolution. The changes in the economic foundation lead sooner or later to the transformation of the whole immense superstructure.[13]

When capitalist production reaches a certain stage of development, the material productive power of society inevitably collides with the capitalistic relations of production. How is it that the relations of production, which had been the developmental form of productive power, are necessarily turned into the opposite, becoming a fetter on productive power? It is this question that Marx's *Capital* seeks to elucidate.

Notes

11. Lenin "On the Question of Dialectics" in *Collected Works* vol. 38 (Moscow: Progress Publishers, 1976) 357-8.

12. Ibid. 358-9.

13. *MECW* vol. 29 (1987), 263.